SATs Skills

Arithmetic Workbook

10–11 years

OXFORD
UNIVERSITY PRESS

Great Clarendon Street, Oxford, OX2 6DP, United Kingdom

Oxford University Press is a department of the University of Oxford.
It furthers the University's objective of excellence in research, scholarship,
and education by publishing worldwide. Oxford is a registered trade mark
of Oxford University Press in the UK and in certain other countries

Text © Sarah Lindsay 2016

Illustrations © Oxford University Press 2016

The moral rights of the author have been asserted

First published in 2016

All rights reserved. No part of this publication may be reproduced, stored
in a retrieval system, or transmitted, in any form or by any means, without
the prior permission in writing of Oxford University Press, or as expressly
permitted by law, by licence or under terms agreed with the appropriate
reprographics rights organization. Enquiries concerning reproduction
outside the scope of the above should be sent to the Rights Department,
Oxford University Press, at the address above.

You must not circulate this work in any other form and you must
impose this same condition on any acquirer

British Library Cataloguing in Publication Data
Data available

978-0-19-274565-1

10 9 8 7

Paper used in the production of this book is a natural, recyclable product
made from wood grown in sustainable forests. The manufacturing process
conforms to the environmental regulations of the country of origin.

Printed in China

Acknowledgements

Cover illustrations: Lo Cole

Although we have made every effort to trace and contact all copyright
holders before publication this has not be possible in all cases. If notified
the publisher will rectify any error or omissions at the earliest opportunity.

Bond SATs Skills Arithmetic 10–11

Unit 1

Number

> 💡 **Helpful Hint**
>
> Each digit in a number has a **place value**. The **place value** tells us the value of that digit, for example the digit 6 in the number 10 658 932 has a value of 6 hundred thousands.
>
> | Ten Millions (TM) | Millions (M) | Hundred Thousands (HTh) | Ten Thousands (TTh) | Thousands (Th) | Hundreds (H) | Tens (T) | Units (U) |
> |---|---|---|---|---|---|---|---|
> | 1 | 0 | 6 | 5 | 8 | 9 | 3 | 2 |

A Answer these questions. You do not need to show your workings out.

1 Which number is bigger, 2 234 653 or 2 234 563?

_____ [1]

2 What is the smallest number you can make using all of these digits?

7 4 8 3 2 7 6

_____ [1]

3 What is the value of the digit 3 in the number 437 785?

_____ [1]

4 Which number is smaller, 9 898 989 or 9 899 898?

_____ [1]

5 Which digit represents the millions in 5 337 289?

_____ [1]

6 Order these numbers, smallest first.

6 755 439   6 756 549   6 755 934   6 756 459

_____ [1]

7 Write this number in digits: seven million, fifty thousand and three.

_____ [1]

8 What is the value of the digit 9 in the number 8 965 453?

_____ [1]

Unit 1

Bond SATs Skills Arithmetic 10–11

> **Helpful Hint**
>
> Numbers can be **rounded** to a nearby value to make them easier to use. You may have to **round** numbers to the nearest ten, hundred, thousand, ten thousand, hundred thousand or million. Remember, if the **next** digit is 4 or less, you **round** down. If the **next** digit is 5 or more, you **round** up.
>
> **Examples:**
>
> 766 **rounded** to the nearest hundred is 800. Hint: Look at the tens digit.
>
> 8164 **rounded** to the nearest thousand is 8000. Hint: Look at the hundreds digit.
>
> 78 942 **rounded** to the nearest ten thousand is 80 000. Hint: Look at the thousands digit.
>
> 823 765 **rounded** to the nearest hundred thousand is 800 000. Hint: Look at the ten thousands digit.
>
> 7 634 742 **rounded** to the nearest million is 8 000 000. Hint: Look at the hundred thousands digit.
>
> 7.664 **rounded** to one decimal place is 7.7. Hint: Look at the hundredths digit.

B Answer these questions. You do not need to show your workings out.

1 Round 67 433 to the nearest hundred.
_____ [1]

2 What is 65 543 rounded to the nearest thousand?
_____ [1]

3 Round 8.343 to the nearest whole number.
_____ [1]

4 What is 6 643 327 rounded to the nearest million?
_____ [1]

5 What is 37.545 rounded to two decimal places?
_____ [1]

6 Round 0.384 276 to the nearest thousandth.
_____ [1]

7 What is 408.927 to the nearest whole number?
_____ [1]

8 What is 289 976 to the nearest hundred?
_____ [1]

Unit 1

Bond SATs Skills Arithmetic 10–11

> **Helpful Hint**
>
> **Negative numbers** are all numbers less than zero.
>
> **Negative numbers** have a minus '–' sign.
>
>
>
> **Positive numbers** are numbers larger than zero. **Positive numbers** sometimes have a '+' sign.

ⓒ Answer these questions. You do not need to show your workings out.

1 Write the next number in this sequence. –15 –10 –5 0 _____ [1]

2 What is the difference between 7 and –2?
   _____ [1]

3 Start with –4. Add 6. What is your answer?
   _____ [1]

4 Write the next number in this sequence. –24 –13 –2 9 _____ [1]

5 What has been subtracted from 2 if the answer is –4?
   _____ [1]

6 If you subtract 6 from –3 what is the answer?
   _____ [1]

7 Which of these pairs of numbers has the greater difference?
   2 and –1   15 and 13
   _____ [1]

8 What is 12 less than 5? _____ [1]

9 Which is bigger, –0.5 or $-\frac{3}{4}$? _____ [1]

10 Add –4 to 6. What is the result? _____ [1]

Unit 1

Bond SATs Skills Arithmetic 10–11

Word problems

(D) Solve these word problems and show your workings out.

1 The closest that the moon comes to Earth is 363 104 km.
   Round this to the nearest thousand kilometres.

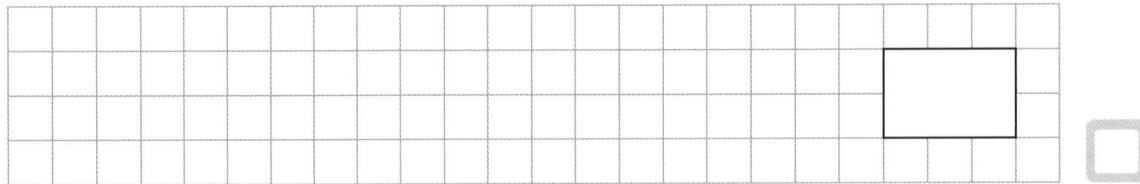

2 During the day, the temperature reached 2°C. At night it dropped by 8°C.
   How cold did it get during the night?

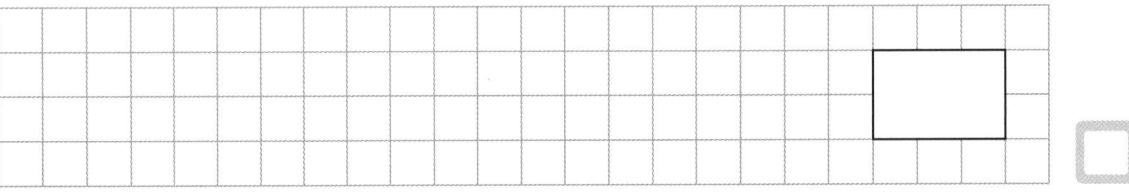

3 Kyle was asked to make the largest number he could using the following digits: 7 3 2 8 1 7 9. Kyle wrote the number correctly.
   What number did he write?

4 The temperature on day 1 rose from −7°C to 5°C. On day 2 it rose from −5°C to 8°C.
   Which day's temperature shows the biggest difference?

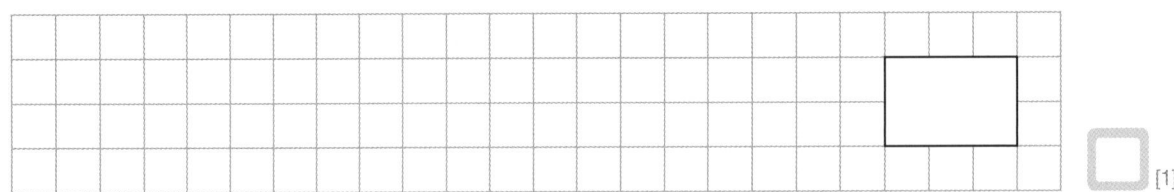

5 The stem of a plant grew 1.34 cm on one day and 1.58 cm on the next day.
   By how much did it grow over the two days, to the nearest millimetre?

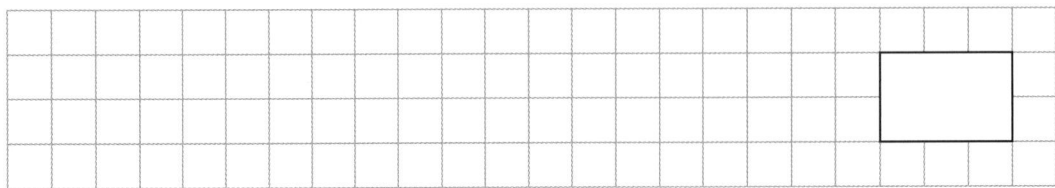

Bond SATs Skills Arithmetic 10–11

Unit 2

Addition and subtraction

> **Helpful Hint**
>
> For some questions, it isn't the answer that we need to find, but part of the question!
>
> **Example:** 100 sweets are shared unequally between 3 friends. 1 friend has 46, another friend has 22 sweets. How many does the third friend have?
>
> 46 + 22 + ? = 100
>
> Using a symbol (?) or letter to represent an unknown number is called **algebra**.

A Answer these questions. You do not need to show your workings out.

1 10 000 − 2300 = _____ [1]

2 35 006 + 8004 = _____ [1]

3 730 + 1900 + 270 = _____ [1]

4 5000 − 3434 = _____ [1]

5 23 100 − ? = 17 800

? = _____ [1]

6 4536 + 2500 − ? = 6000

? = _____ [1]

7 5660 − 2050 + ? = 7260

? = _____ [1]

8 ? + 12 547 = 25 090

? = _____ [1]

Unit 2

Bond SATs Skills Arithmetic 10–11

> 💡 **Helpful Hint**
>
> You can replace any part of a calculation with a letter. **Inverse operations** can be used to re-arrange a calculation to work out which number a letter represents. Remember that the **inverse** of addition is subtraction.
>
> **Example:** $30 + a = 40$ $\qquad\qquad$ $b - 10 = 80$
>
> $\qquad\qquad\;\; 40 - 30 = a$ $\qquad\qquad$ $80 + 10 = 90$
>
> $\qquad\qquad\;\;$ so $a = 10$ $\qquad\qquad\quad$ so $b = 90$

B Answer these questions and show your workings out.

1 $56\,563 + 6529 =$

[1]

2 $67\,865 + a = 145\,318 \qquad a =$

[1]

3 $56\,344 + 12\,876 + 34\,667 =$

[1]

4 $99\,231 + 44\,538 + b = 194\,990 \qquad b =$

[1]

5 $876\,546 + 345\,897 =$

[1]

Bond SATs Skills Arithmetic 10–11

Unit 2

> 💡 **Helpful Hint**
>
> You can split three-number subtraction calculations into two steps.
>
> **Example:** 23 453 – 6732 – 643 = ?
>
> First subtract 6732 from 23 453. 23 453 – 6732 = 16 721
>
> Then subtract the final number from the answer. 16 721 – 643 = 16 078
>
> So, 23 453 – 6732 – 643 = 16 078

(c) Answer these questions and show your workings out.

1 10 453 – 8954 =

[1]

2 77 432 – 38 978 =

[1]

3 66 543 – 7665 – 652 =

[1]

4 34 552 – 29 751 – 98 =

[1]

5 47 565 – 17 304 =

[1]

Unit 2

Bond SATs Skills Arithmetic 10–11

Word problems

(D) Solve these word problems and show your workings out.

1 A political party printed 100 000 leaflets but they only used y leaflets. They had 3277 leaflets left over.

   How many leaflets does y represent?

   [1]

2 Three schools each raised money for a children's charity during one year. School 1 raised £2459, School 2 raised £5548 and School 3 raised £11 407.

   How much money did the schools raise in total?

   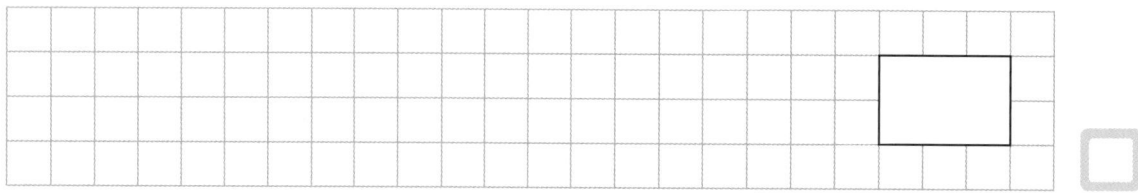
   [1]

3 The Jenkins family drove to Scotland. They left home at 7:10 and arrived at their hotel at 15:27. They stopped for two breaks. One lasted 25 minutes, the other 52 minutes.

   How many hours were they actually driving for?

   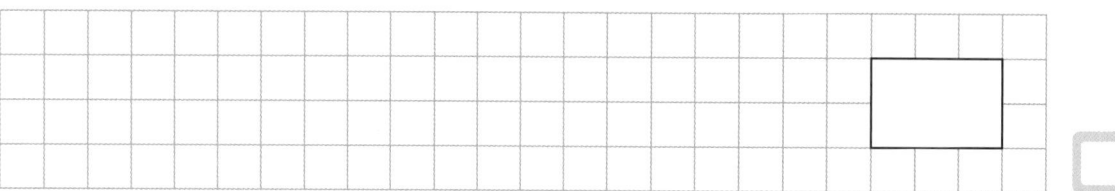
   [1]

4 A supermarket sold x litres of milk in one month. The following month they sold 543 litres more. In two months they sold 22 327 litres of milk.

   What does x stand for?

   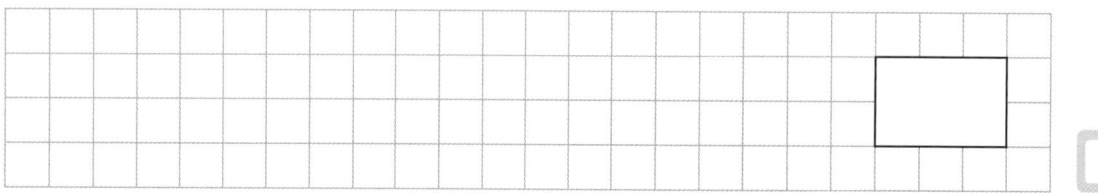
   [1]

Bond SATs Skills Arithmetic 10–11

Unit 3

Multiplication

> **Helpful Hint**
>
> When a number is multiplied by itself and then the result is multiplied by that number again, we say it is **cubed**.
>
> **Example:** $7 \times 7 \times 7$ can be written $7^3 = 343$
>
> The **cube number** of an even number is always even.
>
> The **cube number** of an odd number is always odd.

A Answer these questions. You do not need to show your workings out.

1 $4500 \times 4 =$ _____ [1]

2 $5^3 =$ _____ [1]

3 $798 \times 9 =$ _____ [1]

4 $2337 \times 7 =$ _____ [1]

5 $1889 \times 6 =$ _____ [1]

6 $9^3 =$ _____ [1]

7 $5351 \times 8 =$ _____ [1]

8 $7869 \times 3 =$ _____ [1]

9 $6884 \times 6 =$ _____ [1]

10 $8^3 =$ _____ [1]

Unit 3

Bond SATs Skills Arithmetic 10–11

💡 Helpful Hint

Example showing **long multiplication**:

Multiplying by the tens first

	2	5	3
×		2	7
5₁	0	6	0
1	7₃	7₂	1
6	8₁	3	1

(20 × 253)
(7 × 253)

	2	5	3
×		2	7
1	7₃	7₂	1
5₁	0	6	0
6	8₁	3	1

(7 × 253)
(20 × 253)

Multiplying by the units first

B Answer these questions and show your workings out.

1 641 × 21 =

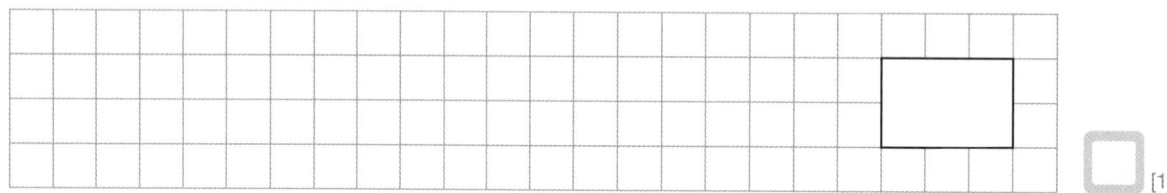
[1]

2 384 × 32 =

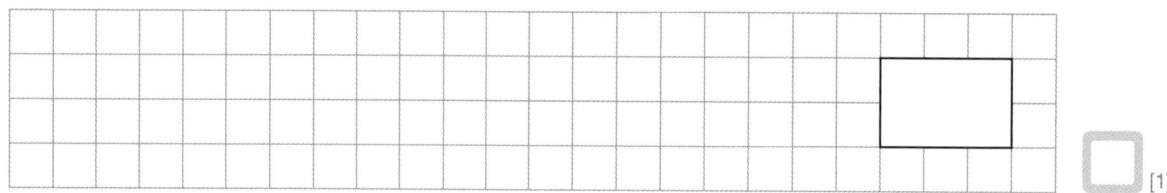
[1]

3 18^3 =

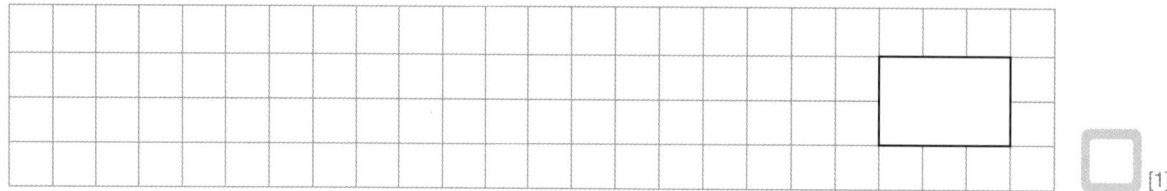
[1]

4 555 × 37 =

[1]

5 25^2 × 46 =

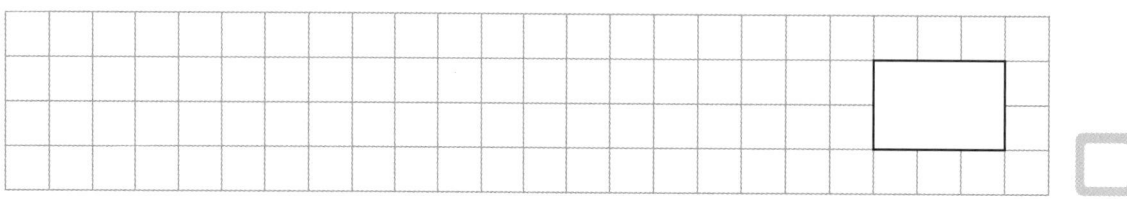
[1]

12

5

Bond SATs Skills Arithmetic 10–11

Unit 3

6 292 × 66 =

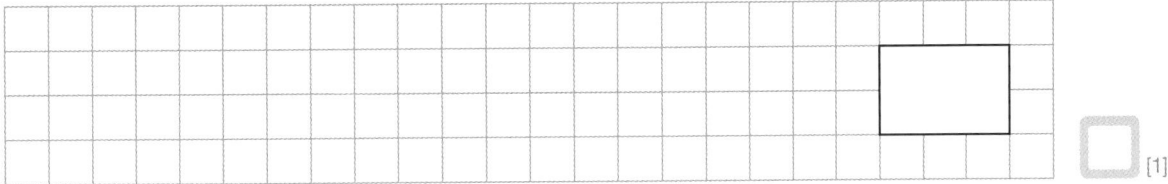

[1]

7 5371 × 14 =

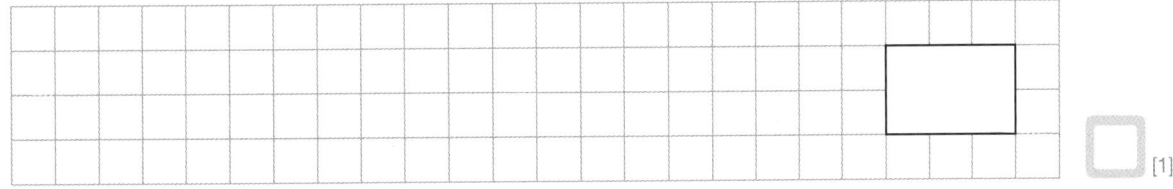

[1]

8 2377 × 36 =

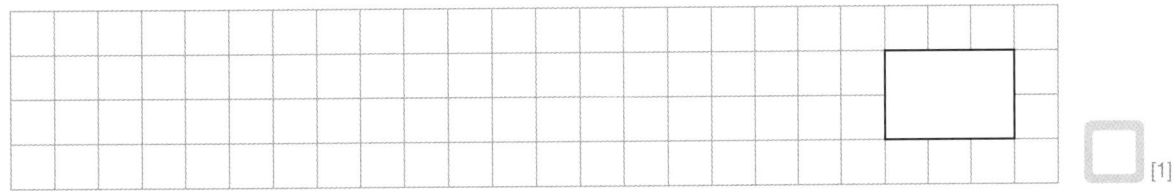

[1]

9 3464 × 42 =

[1]

10 5377 × 46 =

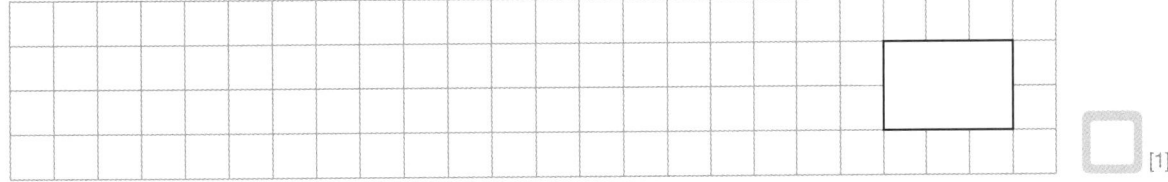

[1]

11 6795 × 27 =

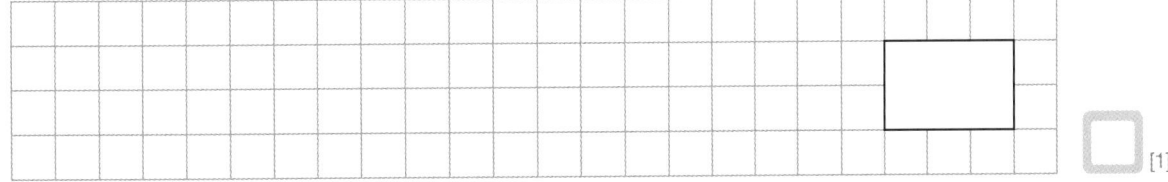

[1]

Unit 3

Bond SATs Skills Arithmetic 10–11

Word problems

> 💡 **Helpful Hint**
> Remember, there are 100 cm in 1 metre and 52 weeks in one year.

ⓒ Solve these word problems and show your workings out.

1 What is the difference between 15^2 and 15^3?

[1]

2 The Leeds Chair Company have an order for 15 chairs. Each chair needs 362 cm of covering fabric.

How many metres of fabric were needed in total?

[1]

3 A long-distance lorry driver drove 1050 miles per week.

How many miles did he drive in total in the first half of the year?

[1]

4 A holiday company organised an exciting trip to South Africa. Each holiday cost £8735. 43 people decided to go on the holiday.

How much money did the holiday company receive?

[1]

Unit 4

Bond SATs Skills Arithmetic 10–11

Division

> **Helpful Hint**
>
> Sometimes a number can't be divided exactly and there is an amount left over. This is known as the **remainder** (r).
>
> **Remainders** can be written as **fractions**.
>
> The **remainder** is the top number (**numerator**) of the **fraction**. The number you are dividing by is the bottom number (**denominator**).
>
> **Example:** $170 \div 6 = 28 \text{ r } 2$ or as a **fraction** $28\frac{2}{6}$ or as a simplified **fraction** $28\frac{1}{3}$

A Answer these questions. You do not need to show your workings out. Write any remainders as fractions in their simplest form.

1 $5680 \div 5 =$ _____ [1]

2 $1155 \div 7 =$ _____ [1]

3 $3048 \div 6 =$ _____ [1]

4 $715 \div 8 =$ _____ [1]

5 $3567 \div 9 =$ _____ [1]

6 $3027 \div 4 =$ _____ [1]

7 $5000 \div 9 =$ _____ [1]

8 $2661 \div 3 =$ _____ [1]

9 $1032 \div 4 =$ _____ [1]

10 $985 \div 7 =$ _____ [1]

Unit 4

Bond SATs Skills Arithmetic 10–11

Helpful Hint

Long division is a way of dividing by bigger numbers. This is because you are unlikely to know the times tables for numbers more than 12. You have to show your workings in a different way. Use multiplication to find the multiples of larger numbers and subtraction to find the **remainders**. Set your work out in columns.

Example of long division:

```
        56 r 2
18)1 0 1 0
   - 9 0 ↓
     1 1 0
   - 1 0 8
         2
```

Step 1 18 doesn't go into 1, so look at the next digit (0). 18 doesn't go into 10, so look at the next digit (1).

	1	8
×		4
	7	2

	1	8
×		5
	9	0

		1	8
×			6
	1	0	8

Step 2 18 goes into 101 five times (18 × 5 = 90). To find the **remainder**, subtract 90 from 101. 101 – 90 = 11

Step 3 Next, bring the 0 down to make 110. 18 goes into 110 six times (18 × 6 = 108). Find the **remainder**: 110 – 108 = 2

So 1010 ÷ 18 = 56 r 2 or 56 $\frac{2}{18}$ or 56 $\frac{1}{9}$. It shows 1010 shared equally between 18 groups gives 56 in each group with 2 left over.

B Answer these questions and show your workings out. Write any remainders as fractions in their simplest form.

1 915 ÷ 14 =

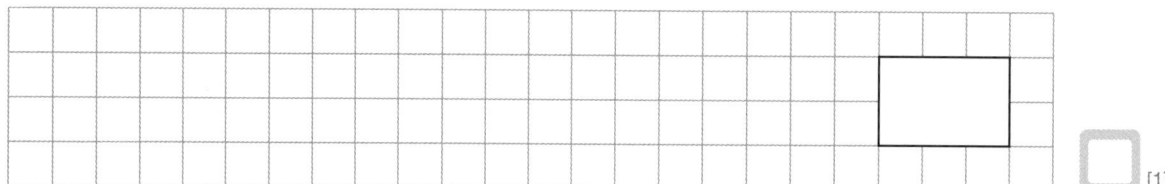

[1]

2 2091 ÷ 17 =

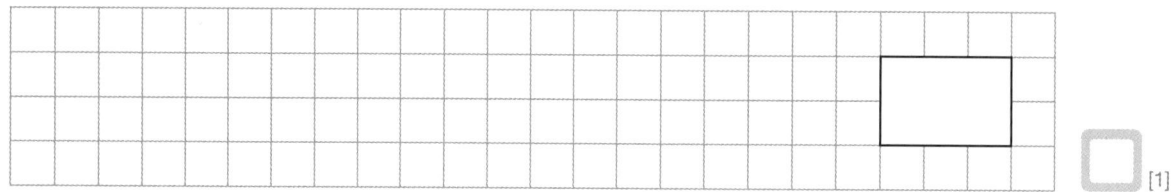

[1]

3 1250 ÷ 26 =

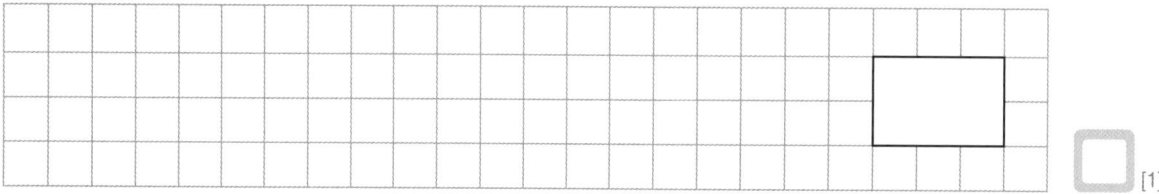

[1]

4 1448 ÷ 32 =

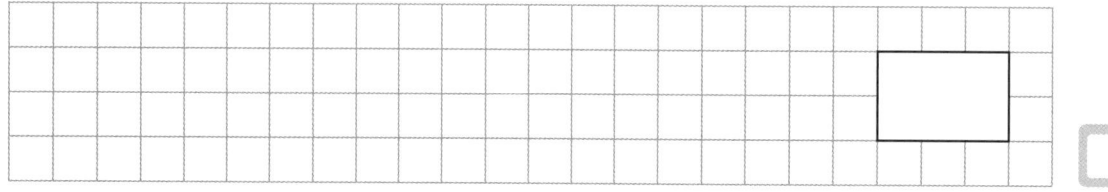

[1]

Unit 4

Bond SATs Skills Arithmetic 10–11

> **Helpful Hint**
> It can also be useful to write down some multiples to get you going by using halving or doubling, which you can write down near the calculation.
>
> **Example:** If dividing by 18, write down 1 × 18 = 18, 2 × 18 = 36, 4 × 18 = 72, 8 × 18 = 144 (doubling) and then 10 × 18 = 180, 5 × 18 = 90 (halving).
>
> This method gives you most of the multiples you need and won't take long to write down.

5 2366 ÷ 23 =

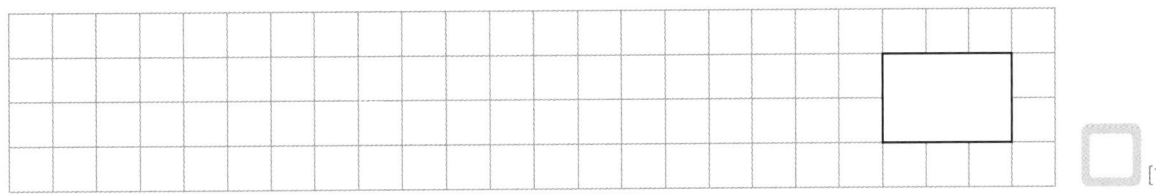

[1]

6 6175 ÷ 42 =

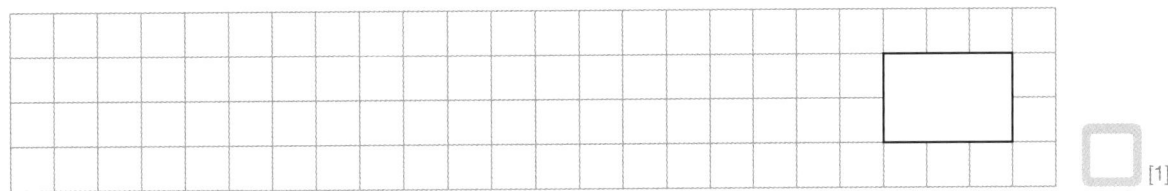

[1]

7 8398 ÷ 39 =

[1]

8 2917 ÷ 19 =

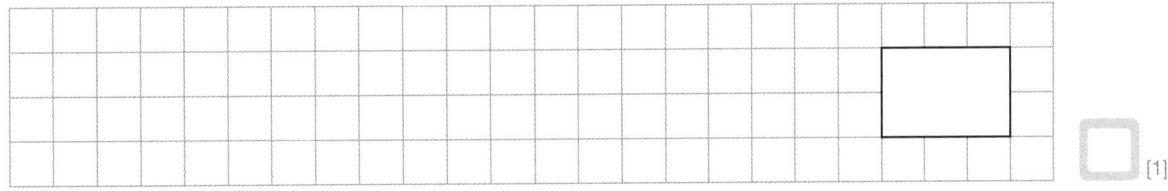

[1]

9 3689 ÷ 34 =

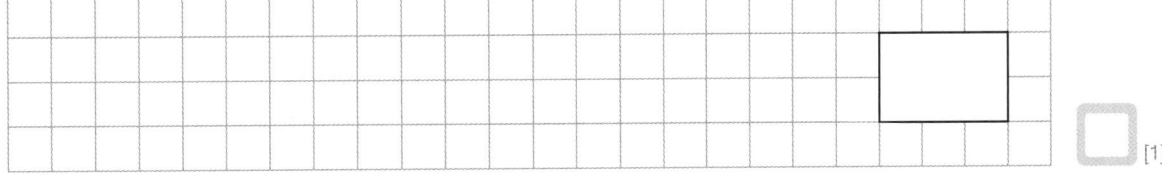

[1]

Unit 4

Bond SATs Skills Arithmetic 10–11

Word problems

> 💡 **Helpful Hint**
> Remember, there are 1000 ml in 1 litre.

ⓒ Solve these word problems and show your workings out.

1 A school has been given an anonymous donation of £4200. It decides to buy notepad computers with the money. Each one costs £75.

   How many notepad computers is the school able to buy?

   [1]

2 1025 lettuces were picked. They were then put into boxes which held a maximum of 15 lettuces each.

   How many boxes were needed for these lettuces?

   [1]

3 Jacob's mum is reading a very long book. It has 1064 pages. She wants to read it in 4 weeks.

   How many pages will she need to read each day to finish the book on time?

   [1]

4 Gita decided to sell hot chocolate drinks to raise money for a local charity. She made 9 litres. Each cup held 200 ml and sold for £1.50. She sold out.

   a How many cups of hot chocolate did she sell?

   b How much money did she raise?

   [2]

Bond SATs Skills Arithmetic 10–11

Unit 5

Multiplication and division

> **Helpful Hint**
>
> You can always check an answer by using its **inverse operation**.
>
> **Example:** $1080 \div 72 = 15$
> $72 \times 15 = 1080$
>
> To use **inverse operations** to check the result of cubing a number, you need to divide your answer twice by the number you cubed.
>
> **Example:** $13^3 = 13 \times 13 \times 13 = 2197$
> $2197 \div 13 \div 13 = 13$

A Answer these questions. You do not need to show your workings out.

1 $2120 \times 6 =$ _____ [1]

2 $276 \div 12 =$ _____ [1]

3 $10^3 =$ _____ [1]

4 $1012 \div 4 =$ _____ [1]

5 $6324 \times 7 =$ _____ [1]

6 $12^3 =$ _____ [1]

7 $3913 \div 7 =$ _____ [1]

8 $3512 \div 8 =$ _____ [1]

9 $9862 \times 7 =$ _____ [1]

10 $15^3 =$ _____ [1]

Unit 5

Bond SATs Skills Arithmetic 10–11

 Helpful Hint
Knowing your times tables well makes multiplication and division calculations much easier. Practise your times tables as often as you can.

B) Answer these questions and show your workings out. Write the answers with remainders as fractions in their simplest form.

1 2375 × 24 =

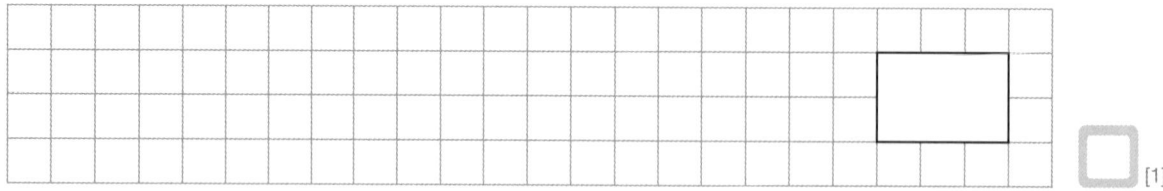

2 1650 ÷ 18 =

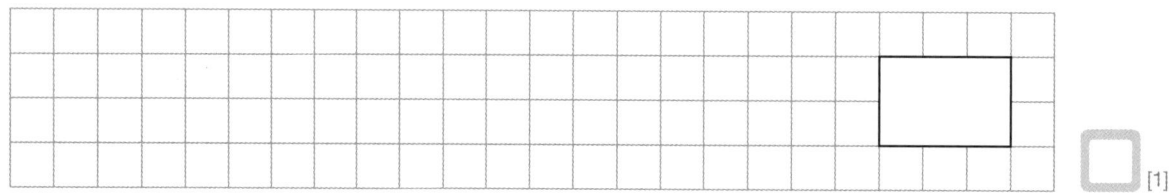

3 3378 × 48 =

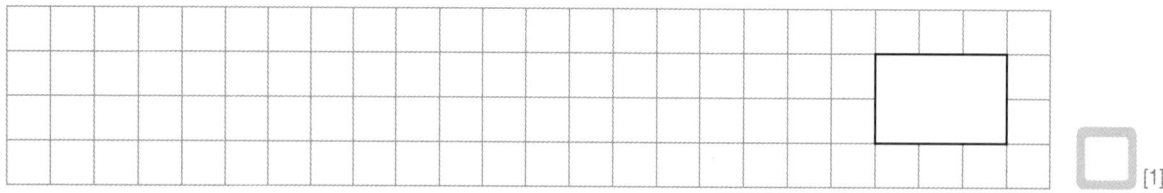

4 9055 ÷ 25 =

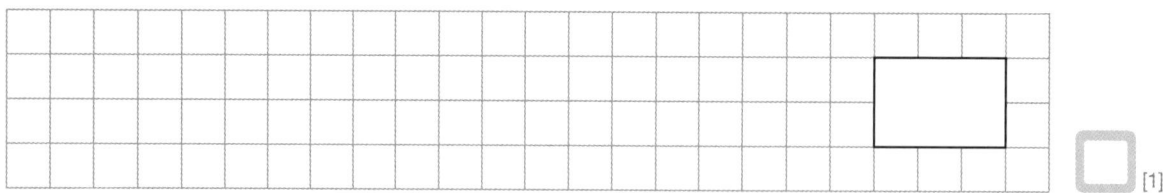

5 8419 × 18 =

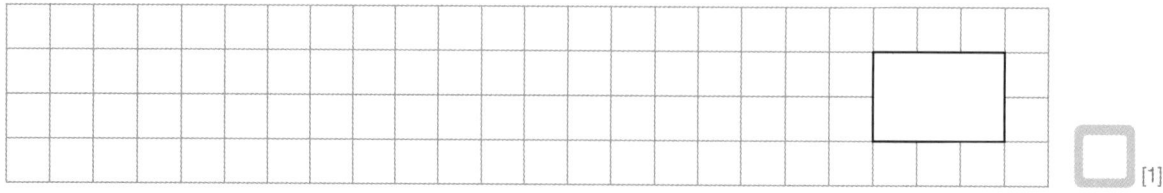

Bond SATs Skills Arithmetic 10–11

Unit 5

6 2300 ÷ 27 =

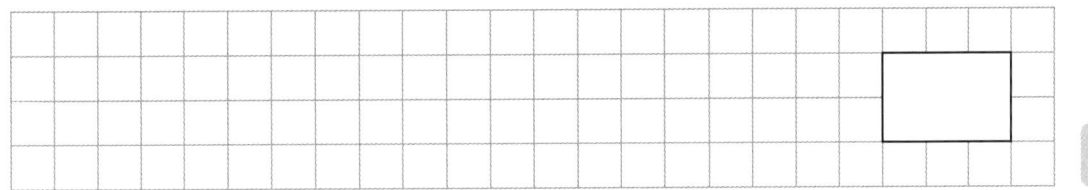
[1]

7 5873 × 81 =

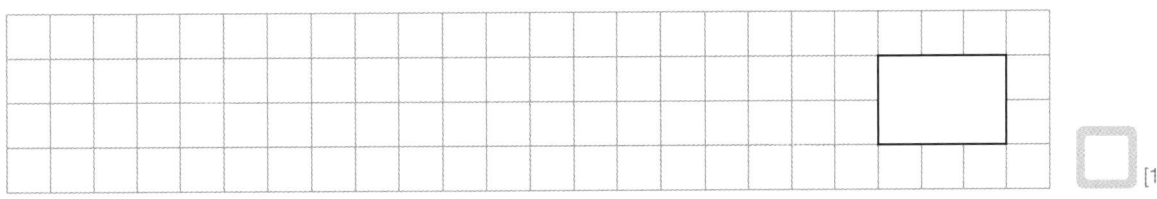
[1]

8 1675 × 59 =

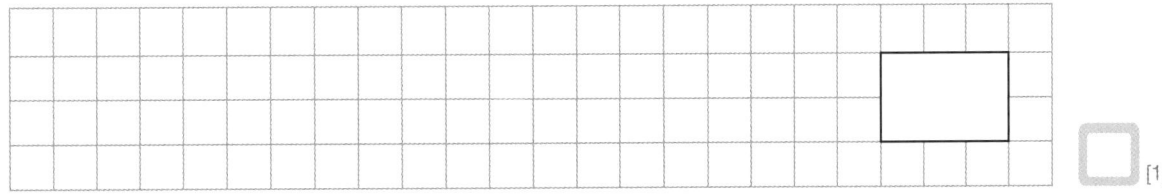
[1]

9 2718 ÷ 36 =

[1]

10 4592 ÷ 56 =

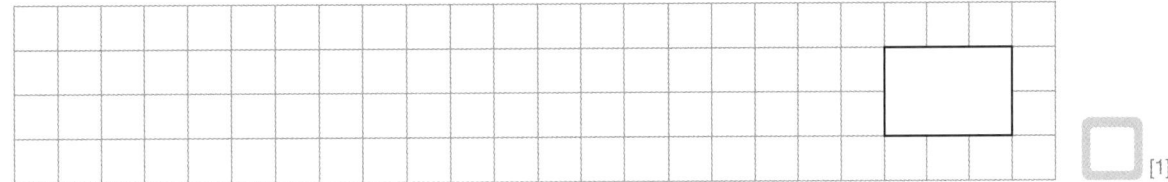
[1]

11 78 008 ÷ 24 =

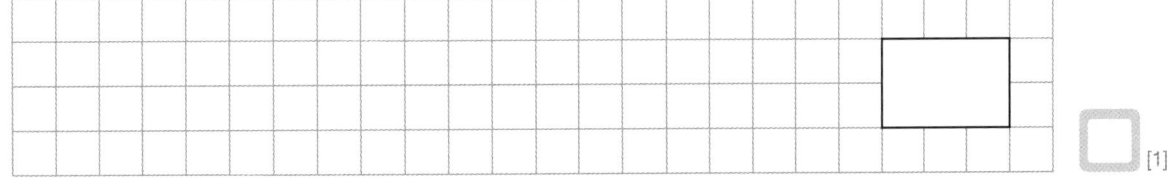
[1]

Unit 5

Bond SATs Skills Arithmetic 10–11

Word problems

ⓒ Answer these questions and show your workings out.

1 Brainstop Primary School has 420 pupils. All the pupils were split into 15 mixed-age teams for sports day.

   How many children were in each team?

   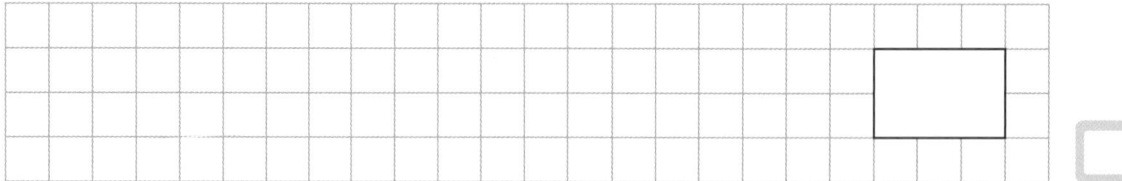

2 If one hairband box holds 1350 hairbands, then 36 boxes hold x hairbands.

   What number does x represent?

3 A chocolatier makes 2455 chocolates each day. In two weeks, working 4 days per week, he makes 19 640 chocolates.

   Is this calculation correct?

4 3712 lambs arrive at the market. In each pen there are 16 lambs and there are y number of pens.

   What is y?

Answers

Bond SATs Skills Arithmetic 10–11

Unit 1 Number

A
1. 2 234 653
2. 2 346 778
3. 30 thousand (30 000)
4. 9 898 989
5. 5
6. **1st** 6 755 439 **3rd** 6 756 459
   2nd 6 755 934 **4th** 6 756 549
7. 7 050 003
8. 9 hundred thousand (900 000)

B
1. 67 400
2. 66 000
3. 8
4. 7 000 000
5. 37.55
6. 0.384
7. 409
8. 290 000

C
1. 5
2. 9
3. 2
4. 20
5. 6
6. −9
7. 2 and −1 (Differences are 3 and 2)
8. −7
9. −0.5
10. 2

Word problems

D
1. 363 000 km
2. −6°C
3. 9 877 321
4. Day 2 (The temperature rose 13°C)
5. 29 mm (2.92 cm = 29.2 mm)

Unit 2 Addition and subtraction

A
1. 7700
2. 43 010
3. 2900
4. 1566
5. ? = 5300
6. ? = 1036
7. ? = 3650
8. ? = 12 543

B
1. 63 092
2. $a = 77\,453$
3. 103 887
4. $b = 51\,221$
5. 1 222 443

C
1. 1499
2. 38 454
3. 58 226
4. 4703
5. 30 261

Word problems

D
1. $y = 96\,723$
2. £19 414
3. 7 hours
4. $x = 10\,892$

Unit 3 Multiplication

A
1. 18 000
2. 125
3. 7182
4. 16 359
5. 11 334
6. 729
7. 42 808
8. 23 607
9. 41 304
10. 512

B
1. 13 461
2. 12 288
3. 5832
4. 20 535
5. 28 750
6. 19 272
7. 75 194
8. 85 572
9. 145 488
10. 247 342
11. 183 465

Word problems

C
1. 3150
2. 54.30 m
3. 27 300 miles
4. £375 605

Unit 4 Division

A
1. 1136
2. 165
3. 508
4. $89\frac{3}{8}$
5. $396\frac{1}{3}$
6. $756\frac{3}{4}$
7. $555\frac{5}{9}$
8. 887
9. 258
10. $140\frac{5}{7}$

B
1. $65\frac{5}{14}$
2. 123
3. $48\frac{1}{13}$
4. $45\frac{1}{4}$
5. $102\frac{20}{23}$
6. $147\frac{1}{42}$
7. $215\frac{1}{3}$
8. $153\frac{10}{19}$
9. $108\frac{1}{2}$

Word problems

C
1. 56 notepad computers
2. 68 boxes with 5 lettuces left over, so 69 boxes needed
3. 38 pages a day
4. **a** 45 cups **b** £67.50

Unit 5 Multiplication and division

A
1. 12 720
2. 23
3. 1000
4. 253
5. 44 268
6. 1728
7. 559
8. 439
9. 69 034
10. 3375

B
1. 57 000
2. $91\frac{2}{3}$
3. 162 144
4. $362\frac{1}{5}$
5. 151 542
6. $85\frac{5}{27}$

7 475 713
8 98 825
9 $75\frac{1}{2}$
10 82
11 $3250\frac{1}{3}$

Word problems

C 1 28 children
2 $x = 48\,600$ hairbands
3 Yes, 19 640 chocolates would be made
4 232 pens

Unit 6 Fractions

A 1 $\frac{3}{5}$
2 $\frac{1}{2}$
3 $5\frac{1}{2}$
4 $\frac{11}{16}$
5 $3\frac{1}{8}$
6 $2\frac{11}{24}$
7 $2\frac{5}{9}$
8 $3\frac{3}{20}$

B 1 $\frac{1}{5}$
2 $\frac{1}{2}$
3 $1\frac{1}{2}$
4 $2\frac{2}{3}$

C 1 $\frac{1}{16}$
2 $\frac{2}{15}$
3 $\frac{8}{15}$
4 $2\frac{2}{3}$
5 $1\frac{6}{7}$

Word problems

D 1 30 pages
2 15 sandwiches
3 6 bulbs
4 7 slices
5 18 friends

Unit 7 Multiplying with decimals

A 1 32
2 28.9
3 5283.7
4 1200
5 67 260
6 40
7 87 999.99
8 0.3786
9 2584

B 1 17.4
2 23.4
3 82.8
4 63.2
5 75.6
6 102.9
7 281.7
8 5.04
9 149.625
10 2.637

Word problems

C 1 6700 g
2 8.8 litres
3 147 km
4 £145

Unit 8 Dividing with decimals

A 1 0.69
2 0.567
3 0.459
4 0.09
5 0.5
6 0.006
7 0.023
8 0.1269
9 2.448
10 0.056
11 18.4634
12 0.000 037

B 1 37.2
2 235.5
3 95.5
4 906.5
5 1904.6
6 85.25

7 245.25
8 65.5
9 187.25
10 339.5
11 10.52

Word problems

(C) **1** 27.5 cm
2 £104.20
3 93.75 g
4 32.5 minutes

Unit 9 Percentages

(A) **1** $\frac{55}{100} = 0.55$
2 $\frac{30}{100} = 0.3$
3 $\frac{70}{100} = 0.7$
4 $\frac{85}{100} = 0.85$
5 $\frac{28}{100} = 0.28$
6 $\frac{79}{100} = 0.79$
7 $\frac{63}{100} = 0.63$
8 $\frac{99}{100} = 0.99$
9 $\frac{48}{100} = 0.48$
10 $\frac{17}{100} = 0.17$

(B) **1** 9
2 66
3 27
4 $16\frac{9}{10}$
5 960
6 11
7 9
8 214
9 312
10 34

Word problems

(C) **1** 6 tonnes
2 60 children
3 21 toffees
4 13 000 bees
5 £410 000. $2\frac{1}{2}$% of £400 000 = £10 000

Unit 10 Test your skills

(A) **1** 0.433
2 $3\frac{1}{2}$
3 2815
4 216
5 3061
6 7125
7 8960
8 $\frac{88}{100}$ or $\frac{44}{50}$ or $\frac{22}{25}$
9 120
10 **1st** 845 797   **3rd** 846 798
   2nd 845 798   **4th** 846 799

(B) **1** 76 958
2 29.6
3 42
4 $1\frac{11}{40}$
5 13
6 $\frac{3}{4}$ of 316 (234 and 237)
7 995
8 199 512
9 $y = 589.5$ or $589\frac{1}{2}$
10 $\frac{3}{75}$ or $\frac{1}{25}$
11 9.889
12 120

Word problems

(C) **1** **a** $a = £6$ **b** $b = £5$
2 −2°C
3 7 pizzas
4 4 questions (80% of 24 = 19.2)

Bond SATs Skills Arithmetic 10–11

Unit 6

Fractions

> **Helpful Hint**
>
> To add and subtract **fractions** with different **denominators**, first write them using the **lowest common multiple** of both **denominators**. This is the smallest number that both **denominators** can divide into.
>
> **Example:** $2\frac{2}{3} + 1\frac{3}{5} = ?$
>
> First turn the **mixed numbers** into **improper fractions**. Multiply the whole number by the **denominator** and then add the **numerator**.
>
> $2 \times 3 = 6 + 2 = \frac{8}{3}$
>
> $1 \times 5 = 5 + 3 = \frac{8}{5}$
>
> Next find the **lowest common multiples** of 3 and 5 (the **denominators**).
>
> Multiples of 3: 3, 6, 9, 12, **15**, 18, 21, …
>
> Multiples of 5: 5, 10, **15**, 20, …
>
> The **lowest common multiple** of 3 and 5 is **15**, so this is the **common denominator**. Now convert both **fractions** to fifteenths. To do this, multiply the **numerator** and the **denominator**.
>
> 3 goes into 15 five times, so multiply the **numerator** and the **denominator** by 5
>
> 5 goes into 15 three times, so multiply the **numerator** and the **denominator** by 3
>
> $8 \times 5 = 40$ $\qquad$ $8 \times 3 = 24$
>
> $3 \times 5 = 15$ $\qquad$ $5 \times 3 = 15$
>
> $\frac{40}{15} + \frac{24}{15} = \frac{64}{15}$
>
> $\frac{64}{15}$ is an **improper fraction**. Divide the **numerator** by the **denominator** to see how many whole numbers ($\frac{15}{15}$) you have.
>
> $64 \div 15 = 4 \text{ r } 4$. So this gives us $4\frac{4}{15}$.

A Answer these questions and write the answers in their simplest form.

1 $\frac{4}{5} - \frac{2}{10} =$ _____ [1]

2 $\frac{3}{9} + \frac{1}{6} =$ _____ [1]

3 $2\frac{2}{3} + 2\frac{5}{6} =$ _____ [1]

4 $\frac{7}{8} - \frac{3}{16} =$ _____ [1]

5 $2\frac{1}{2} + \frac{5}{8} =$ _____ [1]

6 $1\frac{1}{8} + 1\frac{1}{3} =$ _____ [1]

7 $2\frac{1}{6} + \frac{7}{18} =$ _____ [1]

8 $3\frac{2}{5} - \frac{1}{4} =$ _____ [1]

Unit 6

Bond SATs Skills Arithmetic 10–11

> 💡 **Helpful Hint**
>
> Remember, to multiply a **fraction** by a whole number, only multiply the **numerator** by the whole number.
>
> **Example:** $3 \times \frac{3}{10}$ is 3×3 tenths, this is 9 tenths or $\frac{9}{10}$
>
> When multiplying with **mixed numbers**, turn the **mixed number** into an **improper fraction** first.
>
> **Example:** $2\frac{4}{9}$ **Step 1** Multiply the whole number by the **denominator** ($2 \times 9 = 18$)
>
>   **Step 2** Add the result to the **numerator** ($18 + 4 = 22$)
>
>   **Step 3** Replace the **numerator** with the result ($2\frac{4}{9}$ becomes $\frac{22}{9}$)
>
> Then use this to complete the multiplication. Look out for chances to make simpler **equivalent fractions** (here 22 and 11 are **multiples** of 11 and simplify to 2 and 1).
>
> $2\frac{4}{9} \times \frac{5}{11} = \frac{22}{9} \times \frac{5}{11} = \frac{22 \times 5}{9 \times 11} = \frac{2 \times 5}{9 \times 1} = \frac{10}{9} = 1\frac{1}{9}$
>
> If your answer is an **improper fraction**, make sure you convert it to a **mixed number**.
>
> **Example:** $\frac{3}{4} \times 5 = \frac{15}{4}$
>
> $15 \div 4 = 3 \text{ r } 3$ so $\frac{15}{4} = 3\frac{3}{4}$
>
> When multiplying pairs of **fractions**, multiply the **numerators** and then multiply the **denominators**.
>
> **Example:** $\frac{2}{7} \times \frac{3}{4} = \frac{2 \times 3}{7 \times 4} = \frac{6}{28} = \frac{3}{14}$

B Answer these questions and write the answers in their simplest form. Show your workings out.

1 $\frac{3}{5} \times \frac{1}{3} =$

[1]

2 $\frac{5}{6} \times \frac{3}{5} =$

[1]

3 $\frac{2}{3} \times 2\frac{1}{4} =$

[1]

4 $4 \times \frac{2}{3} =$

[1]

4

Bond SATs Skills Arithmetic 10–11

Unit 6

> **Helpful Hint**
>
> To divide **fractions** by whole numbers, multiply the **denominator** by the whole number.
>
> **Example:** $\frac{1}{5} \div 4 = \frac{1}{20}$
>
> $\frac{1}{5 \times 4} = \frac{1}{20}$
>
> To divide a **fraction** by another **fraction**, turn the **fraction** you are dividing by upside down and multiply.
>
> **Example:** $\frac{4}{5} \div \frac{3}{4} = 1\frac{1}{15}$
>
> Turn $\frac{3}{4}$ upside down and multiply:
>
> $\frac{4 \times 4}{5 \times 3} = \frac{16}{15}$
>
> Remember to turn **improper fractions** into **mixed numbers** ($\frac{16}{15}$ becomes $1\frac{1}{15}$).

(c) Answer these questions and write the answers in their simplest form. Show your workings out.

1 $\frac{1}{4} \div 4 =$

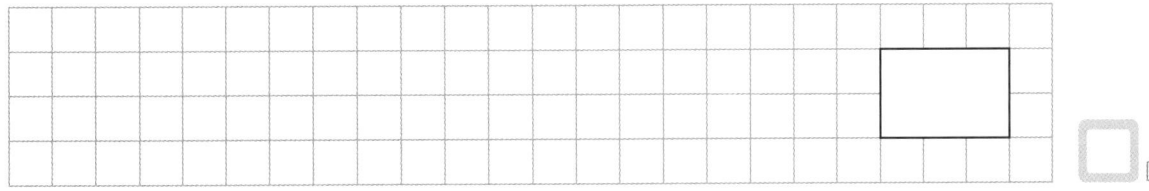

[1]

2 $\frac{4}{5} \div 6 =$

[1]

3 $\frac{2}{5} \div \frac{3}{4} =$

[1]

4 $\frac{5}{8} \div \frac{15}{64} =$

[1]

5 $\frac{2}{3} \div \frac{14}{39} =$

[1]

Unit 6

Bond SATs Skills Arithmetic 10–11

Word problems

(D) Solve these word problems and show your workings out.

1 Maddie's book has 50 pages. In one night she read $\frac{3}{5}$ of it.
   How many pages had she read?

   [1]

2 At a picnic 6 children ate two and a half sandwiches each.
   How many sandwiches were eaten in total?

   [1]

3 A gardener had 36 bulbs to plant. He planted $\frac{5}{6}$ of them in the first hour.
   How many bulbs did he still need to plant?

   [1]

4 A cake was cut into 12 slices. On the first day $\frac{1}{3}$ of the cake was eaten; the following day $\frac{1}{4}$ of the whole cake was eaten.
   How many slices were eaten over the two days?

   [1]

5 I have made 12 large cakes. I want to give my friends $\frac{2}{3}$ of a cake each.
   How many friends can have some cake?

   [1]

Bond SATs Skills Arithmetic 10–11

Unit 7

Multiplying with decimals

> **Helpful Hint**
>
> A **decimal fraction** is a **decimal number** that is less than 1, such as 0.34.
>
> A **mixed decimal** is a **decimal number** that is more than 1, such as 6.34.
>
> When you multiply a number by 10, you move all its digits one place to the left.
>
> When you multiply a number by 100, you move all its digits two places to the left.
>
> When you multiply a number by 1000, you move all its digits three places to the left.
>
> **Example:** 26.4 × 10 = 264
>
> 26.4 × 100 = 2640
>
> 26.4 × 1000 = 26 400
>
> Remember, you do not need to add a .0 to a whole number. Whole numbers are followed by an invisible **decimal point**. For example, write 264 not 264.0.

(A) Answer these questions. You do not need to show your workings out.

1 3.2 × 10 = _____ [1]

2 2.89 × 10 = _____ [1]

3 52.837 × 100 = _____ [1]

4 1.2 × 1000 = _____ [1]

5 672.6 × 100 = _____ [1]

6 0.04 × 1000 = _____ [1]

7 87.999 99 × 1000 = _____ [1]

8 0.003 786 × 100 = _____ [1]

9 25.84 × 100 = _____ [1]

Unit 7

Bond SATs Skills Arithmetic 10–11

 Helpful Hint

When multiplying **decimal numbers**, ignore the **decimal point** and multiply the numbers as if they are whole numbers. Once you have the answer the **decimal point** then needs to be put back in.

See how many decimal places were in the original numbers (4.9 = one decimal place and 3 = no decimal places). Starting from the end of your result (14**7**), count back the same number of decimal places to find where you need to place your **decimal point**.

Example: 4.9 × 3 = ?

49 × 3 = 147

4.9 × 3 = 14.7

B Answer these questions and show your workings out.

1 5.8 × 3 =

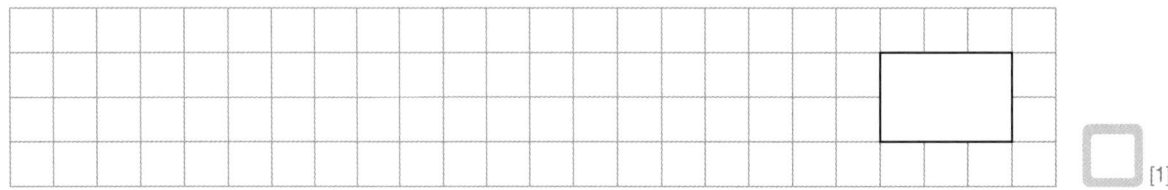

[1]

2 2.6 × 9 =

[1]

3 6.9 × 12 =

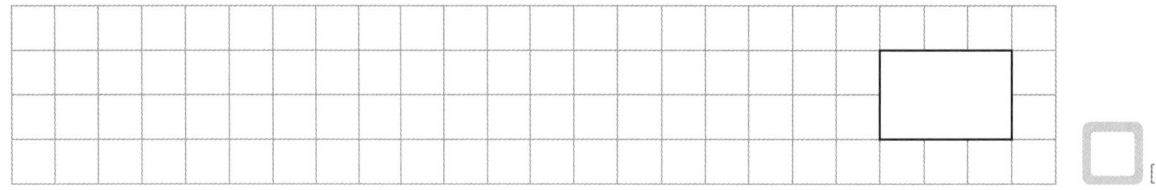

[1]

4 7.9 × 8 =

[1]

Bond SATs Skills Arithmetic 10–11

Unit 7

5 8.4 × 9 =

[1]

6 14.7 × 7 =

[1]

7 31.3 × 9 =

[1]

8 0.7 × 7.2 =

[1]

9 23.75 × 6.3 =

[1]

10 87.9 × 0.03 =

[1]

Unit 7

Bond SATs Skills Arithmetic 10–11

Word problems

 Helpful Hint
Remember, there are 1000 grams in 1 kg.

ⓒ Solve these word problems and show your workings out.

1. Convert 6.7 kg into grams.

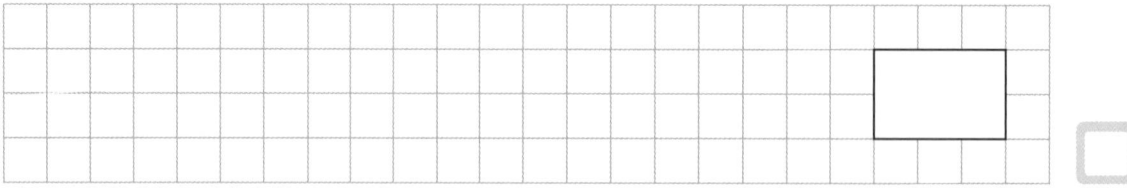

[1]

2. A scientist, as part of an experiment, needed to water five plants. Each plant had to be given 1.76 litres exactly.

   How much water did the scientist need in total?

[1]

3. Six runners each ran 24.5 km while competing in a race.

   What was the total distance run by all of them?

[1]

4. Lana's dad gave her and her three sisters euros to spend during their summer holiday. The euros cost their father £36.25 for each girl.

   How much did it cost him altogether in pounds?

[1]

Unit 8

Bond SATs Skills Arithmetic 10–11

Dividing with decimals

> **Helpful Hint**
>
> Remember, **decimal fractions** are used to show parts of a whole.
>
> When you divide a number by 10, you move its digits one place to the right.
>
> When you divide a number by 100, you move its digits two places to the right.
>
> When you divide a number by 1000, you move its digits three places to the right.
>
> **Example:** 27.9 ÷ 10 = 2.79
>
> 27.9 ÷ 100 = 0.279
>
> 27.9 ÷ 1000 = 0.0279
>
> When working with **decimal numbers** the digits move in the same way.

A Answer these questions. You do not need to show your workings out.

1 6.9 ÷ 10 = _____

2 567 ÷ 1000 = _____

3 45.9 ÷ 100 = _____

4 0.9 ÷ 10 = _____

5 50 ÷ 100 = _____

6 0.6 ÷ 100 = _____

7 23 ÷ 1000 = _____

8 126.9 ÷ 1000 = _____

9 244.8 ÷ 100 = _____

10 56 ÷ 1000 = _____

11 1846.34 ÷ 100 = _____

12 0.0037 ÷ 100 = _____

Unit 8

Bond SATs Skills Arithmetic 10–11

💡 Helpful Hint

Example of long division with decimals: 1356 ÷ 24 = ?

```
       5 6.5
24 )1 3 5 6.0
   -1 2 0 ↓
      1 5 6
     -1 4 4 ↓
         1 2 0
        -1 2 0
             0
```

	2	4
×		4
	9	6

	2	4
×		5
1	2	0

	2	4
×		6
1	4	4

Step 1 24 × 5 = 120

Step 2 135 – 120 = 15

Step 3 Bring down the 6

Step 4 24 × 6 = 144

Step 5 156 – 144 = 12

Step 6 Bring down the 0

Step 7 24 × 5 = 120

So 1356 ÷ 24 = 56.5

B Answer these questions and show your workings out.
Show the remainders as decimals.

1 186 ÷ 5 =

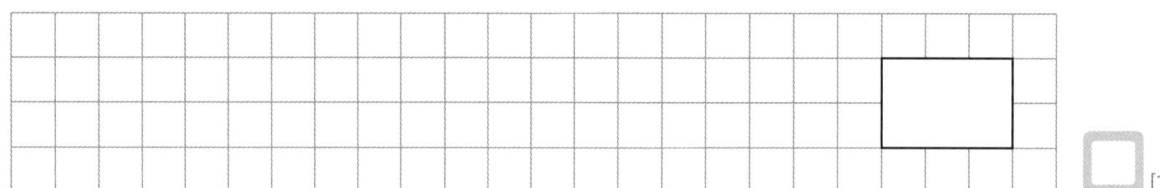

[1]

2 942 ÷ 4 =

[1]

3 573 ÷ 6 =

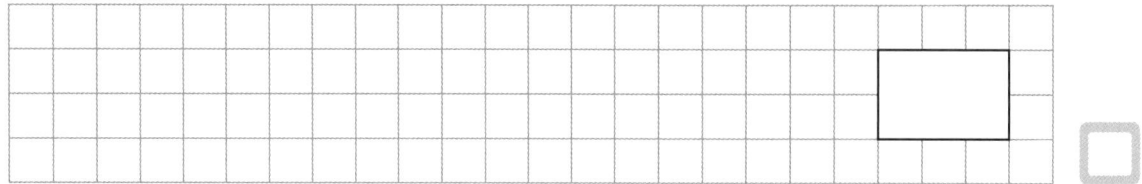

[1]

4 7252 ÷ 8 =

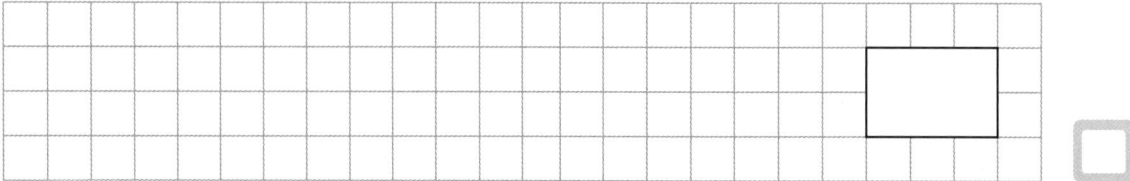

[1]

Bond SATs Skills Arithmetic 10–11

Unit 8

5 9523 ÷ 5 =

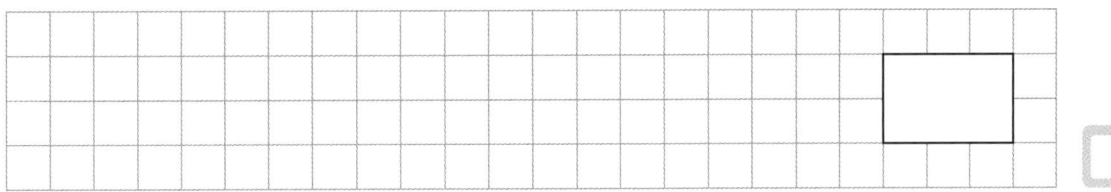
[1]

6 1023 ÷ 12 =

[1]

7 1962 ÷ 8 =

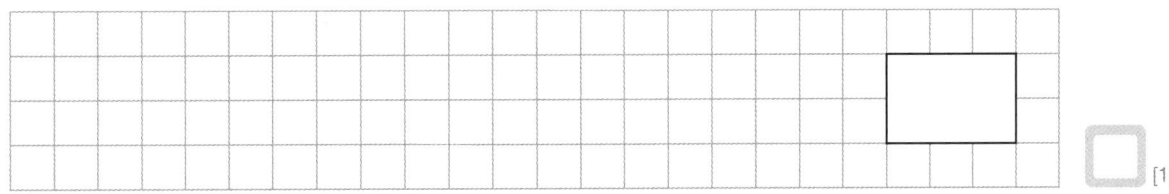
[1]

8 1441 ÷ 22 =

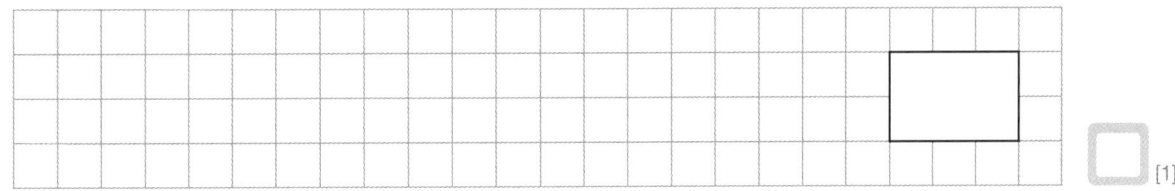
[1]

9 3745 ÷ 20 =

[1]

10 5432 ÷ 16 =

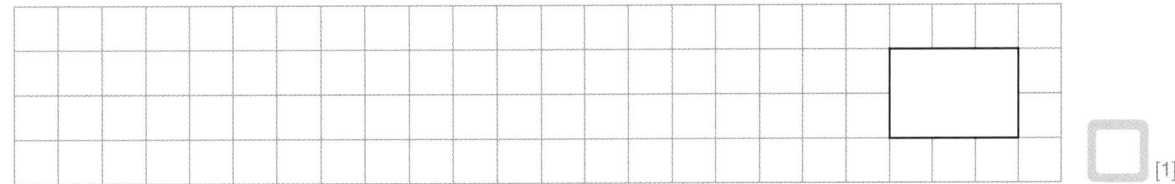
[1]

11 263 ÷ 25 =

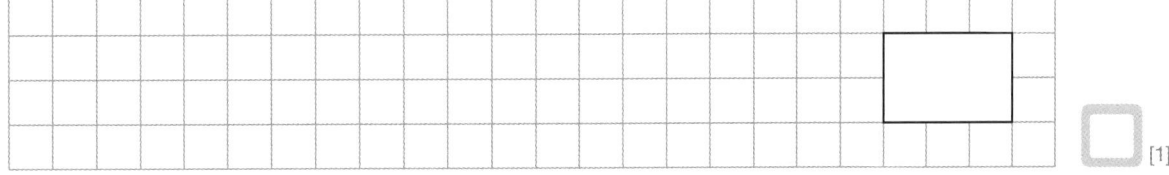
[1]

Unit 8

Bond SATs Skills Arithmetic 10–11

Word problems

> **Helpful Hint**
> Remember, **decimal points** are used in money to separate the pounds from the pence.
> **Example:** £47.98

ⓒ Solve these word problems and show your workings out.

1 A length of ribbon is cut into equal pieces and used to tie back eight dancers' hair. The ribbon is 220 cm.

   How long will each ribbon be?

   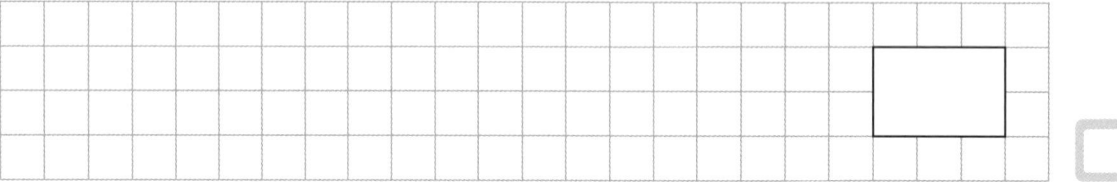

   [1]

2 The Rawlings children were given £521 by their Grandma for Christmas. The five children shared the money equally.

   How much money did they each receive?

   [1]

3 A 750 g piece of chocolate was shared equally between eight children.
   How many grams of chocolate did they each get?

   [1]

4 Ani completed her 5 km run in 1950 seconds.
   How many minutes did it take her?

   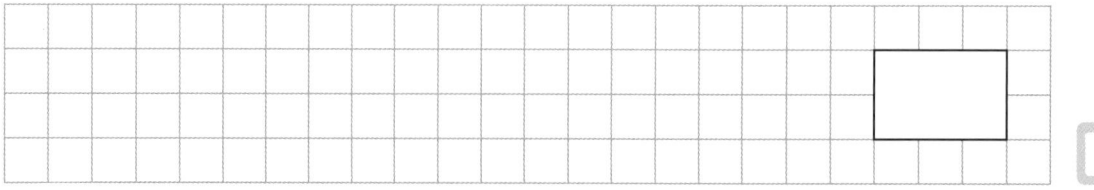

   [1]

Bond SATs Skills Arithmetic 10–11

Unit 9

Percentages

> **Helpful Hint**
> A **percentage** is a number or amount in each hundred.
> **Example:** 35% means 35 parts out of each 100.
> As a **fraction** it is written: $\frac{35}{100}$
> As a **decimal number** it is written: 0.35

A Write these percentages as fractions and decimals.

1 55% = $\frac{}{100}$ = 0. _____ [1]

2 30% = $\frac{}{100}$ = 0. _____ [1]

3 70% = $\frac{}{100}$ = 0. _____ [1]

4 85% = $\frac{}{100}$ = 0. _____ [1]

5 28% = $\frac{}{100}$ = 0. _____ [1]

6 79% = $\frac{}{100}$ = 0. _____ [1]

7 63% = $\frac{}{100}$ = 0. _____ [1]

8 99% = $\frac{}{100}$ = 0. _____ [1]

9 48% = $\frac{}{100}$ = 0. _____ [1]

10 17% = $\frac{}{100}$ = 0. _____ [1]

Unit 9

Bond SATs Skills Arithmetic 10–11

> **Helpful Hint**
>
> To find the **percentage** of a number you can change the **percentage** and number into a **fraction** and multiply. Look out for chances to make simpler **equivalent fractions**.
>
> Here $\frac{25}{100}$ simplifies to $\frac{1}{4}$. 40 and 4 are **multiples** of 4 and simplify to 10 and 1.
>
> **Example:** 25% of 40 = ?
>
> Remember, 25% means 25 **over** or **out of** 100.
>
> $\frac{25}{100} \times \frac{40}{1} = \frac{1}{4} \times \frac{40}{1} = \frac{1 \times 40}{4 \times 1}$
>
> $\frac{1 \times 10}{1 \times 1} = \frac{10}{1} = 10$ so 25% of 40 = 10

B Answer these questions and show your workings out.

1 30% of 30 =

[1]

2 55% of 120 =

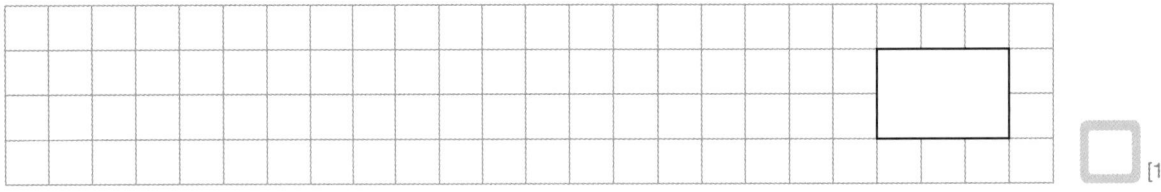
[1]

3 18% of 150 =

[1]

4 65% of 26 =

[1]

5 96% of 1000 =

[1]

Bond SATs Skills Arithmetic 10–11

Unit 9

> **Helpful Hint**
>
> 1% = $\frac{1}{100}$ so to find 1% of a number, you need to divide it by 100.
>
> **Example:** 1% of 65 = 65 ÷ 100 = 0.65
>
> 10% = $\frac{10}{100}$ = $\frac{1}{10}$ so to find 10% of a number, you need to divide it by 10.
>
> **Example:** 10% of 65 = 65 ÷ 10 = 6.5
>
> 20% = $\frac{20}{100}$ = $\frac{1}{5}$ so to find 20% of a number, you need to divide it by 5.
>
> **Example:** 20% of 65 = 65 ÷ 5 = 13
>
> You could also double the answer to 10% to find 20% of something, or halve the 10% to find 5% of something.

6 25% of 44 =

7 15% of 60 =

8 50% of 428 =

9 75% of 416 =

10 5% of 680 =

Unit 9

Bond SATs Skills Arithmetic 10–11

Word problems

c) Solve these word problems and show your workings out.

1 A farmer harvested his crop of corn. He had a total of 24 tonnes of corn. The first lorry took 25% of the corn to market.

   How much corn did the lorry take?

   [1]

2 There were 80 children in Year 6 at Bardfield School. 75% took part in the school play.

   How many children were in the play?

   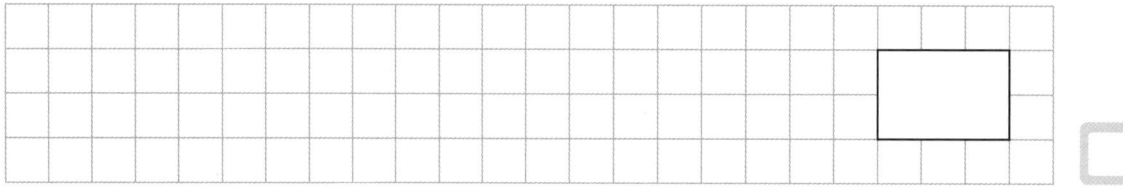
   [1]

3 In a bag of 60 sweets, 35% were toffees.

   How many toffees were there?

   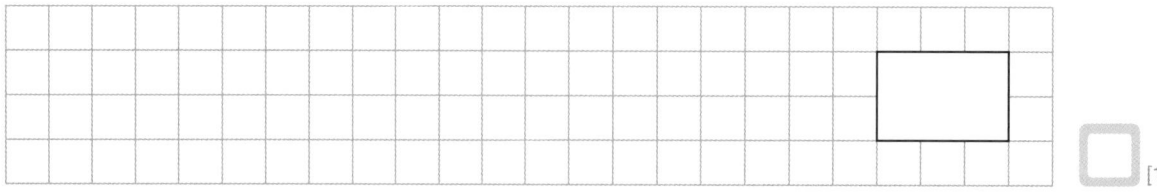
   [1]

4 20 000 bees lived in a hive. During the day 65% went out to find pollen.

   How many bees left the hive to find pollen?

   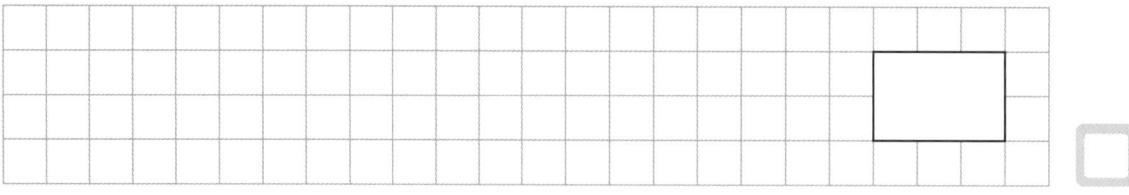
   [1]

5 A company made £400 000 profit in one year. In the second year their profit had increased by $2\frac{1}{2}$%.

   How much was their profit in the second year?

   [1]

Bond SATs Skills Arithmetic 10–11

Unit 10

Test your skills

> **Helpful Hint**
> The questions in this unit are all mixed calculations. Time yourself to see how well you can do. Look back at the other units if you need to refresh your memory.

A Answer these questions. You do not need to show your workings out.

1 $433 \div 1000 =$ _____ [1]

2 $1\frac{1}{3} + 2\frac{1}{6} =$ _____ [1]

3 $563 \times 5 =$ _____ [1]

4 $6^3 =$ _____ [1]

5 $4500 - 1439 =$ _____ [1]

6 $5600 + 1450 + 75 =$ _____ [1]

7 $89.6 \times 100 =$ _____ [1]

8 Write 88% as a fraction: _____ [1]

9 What is 40% of 300? _____ [1]

10 Order these numbers, smallest first. 845 798   846 799   846 798   845 797

_____ [1]

Unit 10

Bond SATs Skills Arithmetic 10–11

B Answer these questions and show your workings out.

1 67 004 + 9954 =

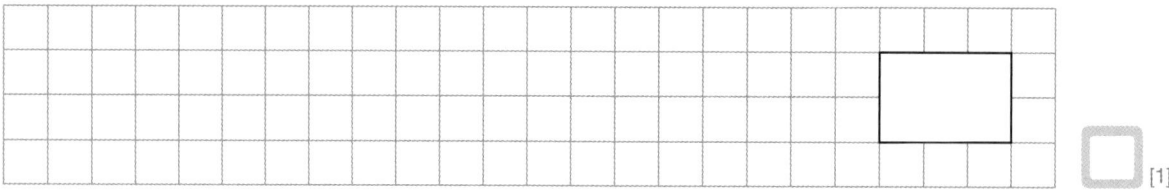

[1]

2 3.7 × 8 =

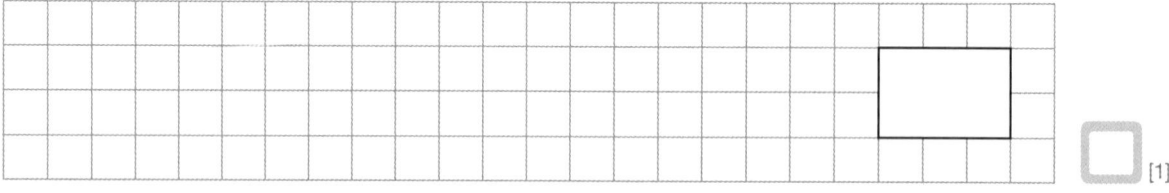

[1]

3 70% of 60 =

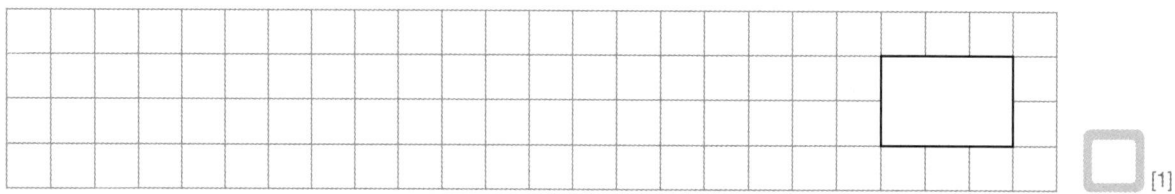

[1]

4 $\frac{3}{5} + 2\frac{1}{8} =$

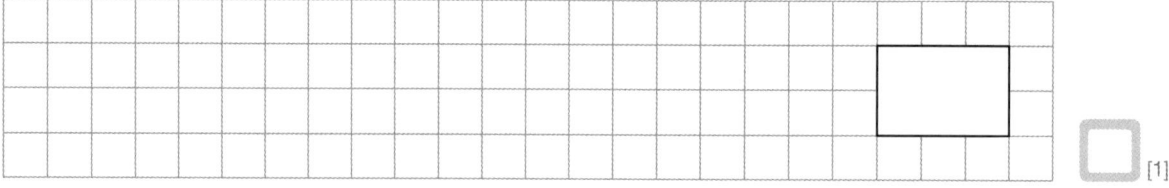

[1]

5 What is the difference between 9 and −4?

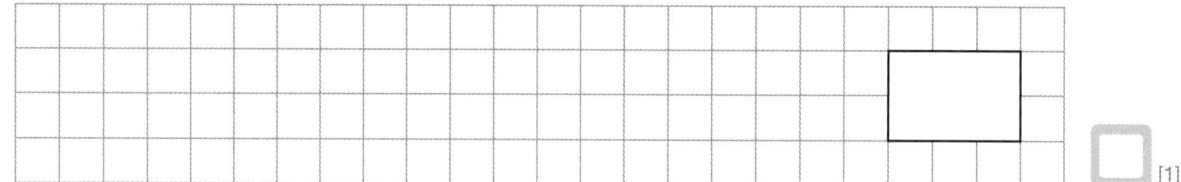

[1]

6 Which is greater, 72% of 325 or $\frac{3}{4}$ of 316?

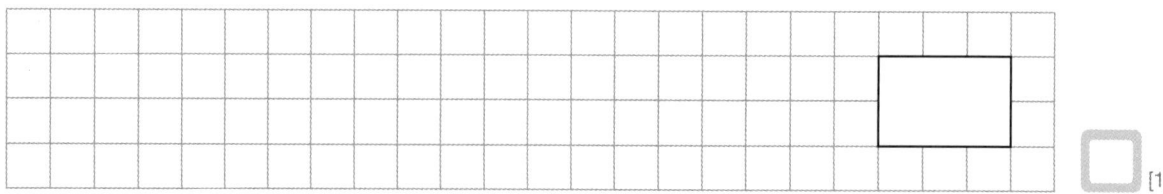

[1]

Unit 10

Bond SATs Skills Arithmetic 10–11

7 64 354 − a = 63 359   a =

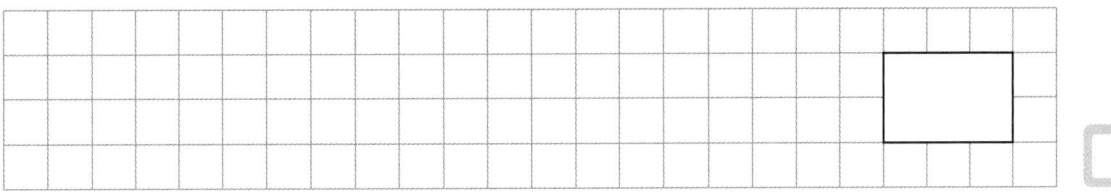
[1]

8 5542 × 36 =

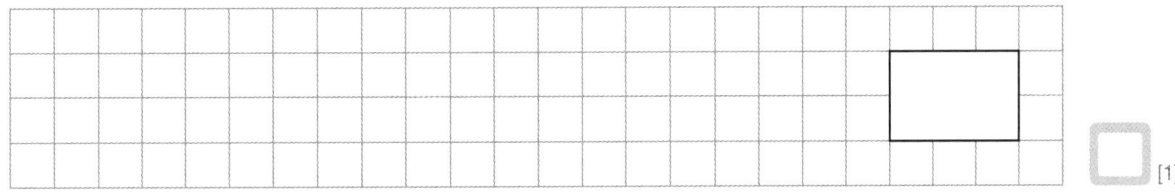
[1]

9 2358 ÷ y = 4   y =

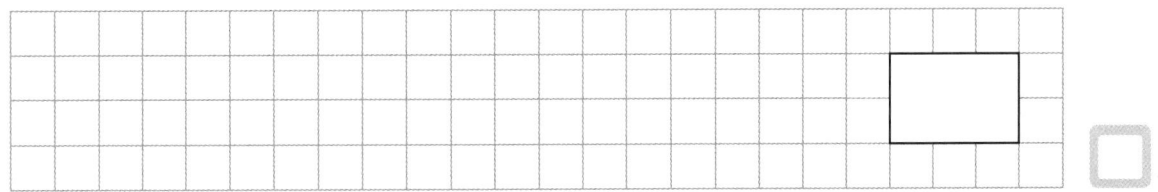
[1]

10 $\frac{3}{5}$ ÷ 15 =

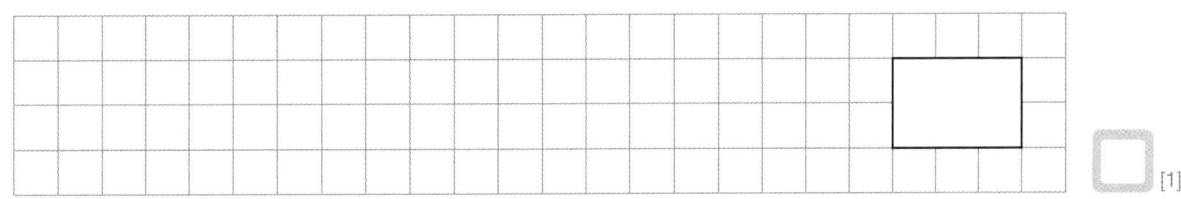
[1]

11 988.9 ÷ 100 =

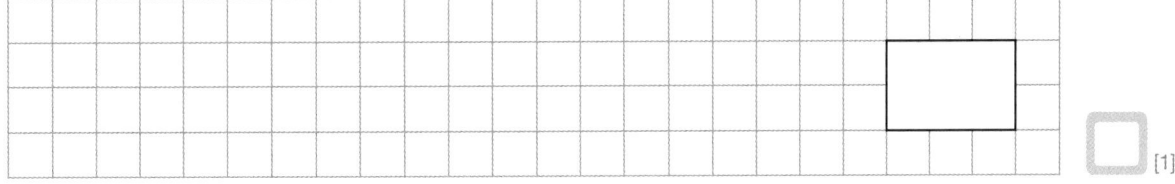
[1]

12 6840 ÷ 57 =

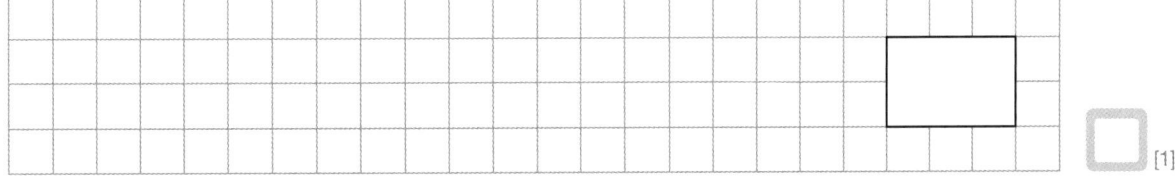
[1]

Unit 10

Bond SATs Skills Arithmetic 10–11

Word problems

ⓒ Solve these word problems and show your workings out.

1. Meena bought seven gifts, one for each of the teachers that had taught her during primary school.

   She bought three gifts at one price (a) and four gifts at a lower price (b). The difference in price between price a and price b is £1. Meena paid a total of £38.
   The amount she spent could be written as $a + a + a + b + b + b + b = £38$.

   a What does a represent?
   b What does b represent?

   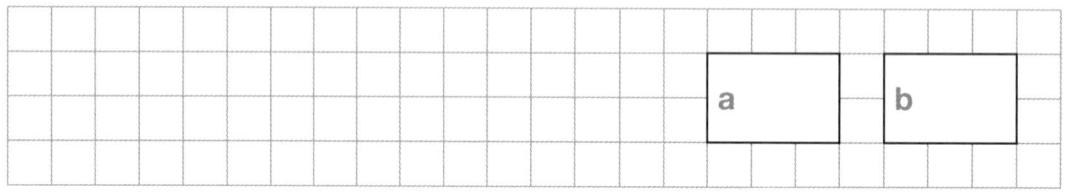

   [2]

2. On a chilly winter's morning the temperature was −8°C. By 4 o'clock that afternoon the temperature had risen by six degrees.

   What was the temperature at 4 o'clock?

   [1]

3. There will be eight children at Jasmine's party. Her father thinks the children will eat about $\frac{5}{6}$ of a pizza each.

   How many pizzas does he need to buy?

   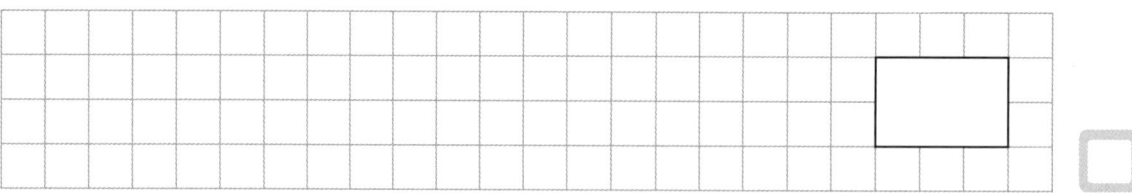

   [1]

4. Jodi has to do a test with 24 questions. She must get at least 80% correct to get a Grade A. What is the most number of questions she could get wrong and still have a Grade A?

   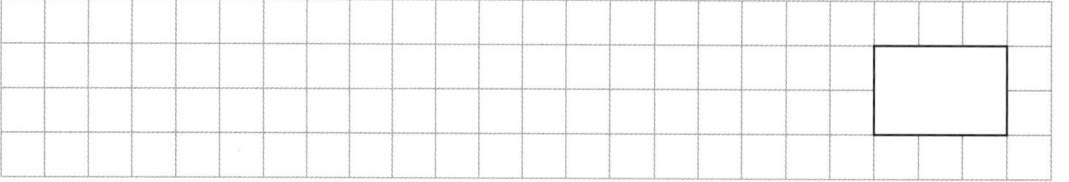

   [1]

Key words

Algebra using letters to represent numbers

Common denominator a multiple of the denominators of two or more fractions, for example $\frac{2}{3} + \frac{3}{4} = \frac{8}{12} + \frac{9}{12}$. The common denominator here is 12 (12 is the lowest common multiple of 3 and 4)

Cube number a number that is the result of a number being multiplied by itself then multiplied by itself again

Decimal fraction a fraction that has 10, 100 or 1000 as the denominator that can then be placed on the decimal system using a decimal point, for example $\frac{40}{100} = 0.40$ (40 hundredths)

Decimal number a number that uses the decimal point

Decimal point a decimal point comes between the units and the tenths. It separates whole numbers from fractions

Denominator the bottom number of a fraction is called the denominator. It tells you how many equal parts the shape or amount has been divided into

Equivalent fraction a fraction which has the same value as another fraction but has been written with different numbers, for example $\frac{35}{40}$ is an equivalent fraction to $\frac{7}{8}$

Fraction a fraction is a part of one whole thing

Improper fraction a fraction where the numerator is larger than the denominator

Inverse operation this reverses the effect of another operation, for example addition is the inverse of subtraction and multiplication is the inverse of division. $2 + 3 = 5$ and $5 - 3 = 2$, $6 \times 2 = 12$ and $12 \div 2 = 6$. We can use inverse operations to check our answers

Long division a column method of working out a division

Long multiplication a column method of working out a multiplication

Lowest common multiple (LCM) the lowest multiple of two or more numbers, for example the LCM of 2 and 3 is 6 (multiples of 2 = 2, 4, 6, 8, ...) (multiples of 3 = 3, 6, 9, ...)

Mixed decimal a mixed decimal has whole numbers before the decimal point and numbers after the decimal point, for example 7.23

Mixed number a number shown with a whole number and a fraction, for example $2\frac{1}{2}$

Multiple a multiple is one number multiplied by another, for example the first three multiples of 3 are 3, 6 and 9

Negative numbers negative numbers are numbers less than zero

Number sequence a series of numbers that form a pattern, for example 10, 11, 12, 13 or 10, 20, 30, 40

Numerator the top number of a fraction is called the numerator. It tells you how many parts of the total amount have been taken, for example **two** thirds or **one** half

Percentage (%) per cent means 'out of 100', for example 52% is 52 out of 100

Place value tells you the value of a digit in a number, for example 2 units, 2 tenths or 2 hundreds

Remainder the amount left over when one number does not divide exactly into another number

Rounding if the digit to the right of the digit we are rounding to is 0, 1, 2, 3 or 4 we round the number down. If it is 5 or above, we round the number up

Simplest form reducing the numbers in a fraction or ratio until they can no longer be divided by a common factor

Progress chart

Bond SATs Skills Arithmetic 10–11

How did you do? Fill in your score. Shade the matching boxes so that you can see how well you are doing in the different units.

Unit 1, p3 Score: __ / 8	
Unit 1, p4 Score: __ / 8	
Unit 1, p5 Score: __ / 10	
Unit 1, p6 Score: __ / 5	
Unit 2, p7 Score: __ / 8	
Unit 2, p8 Score: __ / 5	
Unit 2, p9 Score: __ / 5	
Unit 2, p10 Score: __ / 4	
Unit 3, p11 Score: __ / 10	
Unit 3, p12 Score: __ / 5	
Unit 3, p13 Score: __ / 6	
Unit 3, p14 Score __ / 4	
Unit 4, p15 Score: __ / 10	
Unit 4, p16 Score: __ / 4	
Unit 4, p17 Score: __ / 5	
Unit 4, p18 Score: __ / 5	
Unit 5, p19 Score: __ / 10	
Unit 5, p20 Score: __ / 5	
Unit 5, p21 Score: __ / 6	
Unit 5, p22 Score: __ / 4	
Unit 6, p27 Score: __ / 8	
Unit 6, p28 Score: __ / 4	
Unit 6, p29 Score: __ / 5	
Unit 6, p30 Score: __ / 5	
Unit 7, p31 Score: __ / 9	
Unit 7, p32 Score: __ / 4	
Unit 7, p33 Score: __ / 6	
Unit 7, p34 Score: __ / 4	
Unit 8, p35 Score: __ / 12	
Unit 8, p36 Score: __ / 4	
Unit 8, p37 Score: __ / 7	
Unit 8, p38 Score: __ / /4	
Unit 9, p39 Score: __ / 10	
Unit 9, p40 Score: __ / 5	
Unit 9, p41 Score: __ / 5	
Unit 9, p42 Score: __ / 5	
Unit 10, p43 Score: __ / 10	
Unit 10, p44 Score: __ / 6	
Unit 10, p45 Score: __ / 6	
Unit 10, p46 Score: __ / 5	